Kelizângela N. Albuquerque
Cristina R. B. Silva
José Thales P. Ferreira

Students' perception of the geography of Santana do Mundaú - Alagoas

Kelizângela N. Albuquerque
Cristina R. B. Silva
José Thales P. Ferreira

Students' perception of the geography of Santana do Mundaú - Alagoas

Geography

ScienciaScripts

Cover image: www.ingimage.com

This book is a translation from the original published under ISBN 978-3-330-76688-4.

Publisher:
Sciencia Scripts
is a trademark of
Dodo Books Indian Ocean Ltd. and OmniScriptum S.R.L publishing group

120 High Road, East Finchley, London, N2 9ED, United Kingdom
Str. Armeneasca 28/1, office 1, Chisinau MD-2012, Republic of Moldova, Europe
Managing Directors: Ieva Konstantinova, Victoria Ursu
info@omniscriptum.com

Printed at: see last page
ISBN: 978-620-8-54387-7

CONTENTS

CHAPTER 1

INTRODUCTION

Bearing in mind the role of geography teaching in understanding the problems of today's world, and understanding that these problems almost always have a strong territorial component, we investigated the approach to physical geography themes in high school geography classes, in an attempt to present and discuss the students' perception of the landforms that surround them.

In order to gain a geographical understanding of the world today, with its different spatialisations and contexts, it is necessary to establish bridges between information and facts, according to the appropriate scales of approach and understanding, starting from the immediate space, the everyday, and going up to the most diverse spatial instances, (OLIVEIRA and MIRANDA, 2010). As a result, there is an increasing need to teach a Geography that is coherent with global reality, making students think rationally and critically, positioning themselves in relation to the various issues that will be imposed on them during their journey.

The nature of the research carried out in this study was based on a survey to find out whether the teachers at the Manuel de Matos School in the municipality of Santana do Mundaú/ AL used the local area in their lessons, comprising the totality of teacher-student involvement in educational issues related to the municipality.

Geographically, the municipality of Santana do Mundaú is located in a region surrounded by mountains, which is peculiar to the surrounding municipalities. With these unconventional characteristics in mind, we asked ourselves how the citizens, especially the students, perceived this region, how their knowledge was in relation to what they experienced there. We therefore had the general objective of understanding the totality of teacher-student involvement in educational issues related to the municipality, in order to draw conclusions. We began the research by analysing the perception of primary school students at the Manoel de Matos State School regarding local geography, as well as probing how teachers approach such topics and how the contents of the textbook are related to the type of structure presented by the municipality, by applying questionnaires with questions directed at students in the afternoon (90 questionnaires) and evening (45 questionnaires) shifts (model in the appendix). The questions were structured to check the students' ability to identify geographical features present in the municipality of Santana do Mundaú- AL and the region.

It is necessary for learning that the teacher in the classroom articulates the content and the environment that surrounds the student, thus facilitating their understanding. As the chosen city is rich in places to be explored, we asked whether teachers use these means to facilitate the student's conception of the content taught in the classroom, which is almost always based on the textbook. In order to reach a conclusion, we drew up a questionnaire made up of 14 questions relating to physical geography and the students' opinion of the subject, which formed the basis of the research. From the students' answers, some conclusions can be drawn that will be discussed later.

CHAPTER 2

GEOGRAPHY TEACHING IN BRAZIL

Geography has historically travelled several paths, mainly as a school subject, undergoing significant changes in its form. The first teachers trained to work in secondary education bore the marks of classical Geography, strongly influenced by Vidal de La Blache, where according to Cassab (2009) it was translated into the descriptive study of humanised and natural landscapes, marked by the delineation and memorisation of elements, with the student being responsible for relating natural and social facts, making their own analogies, generalisations and syntheses. The cultural and political criteria of society were omitted. Man and nature were studied separately.

This Geography had a purely informative core, and after a few years it was criticised, giving way to the new Geography from a theoretical-quantitative perspective, with nature focused on a geometric and hierarchical space. Both forms of these geographies boiled down to memorising content, without thinking about or interpreting the lived space, the surrounding reality.

New ways of understanding space, landscape, territory and place are emerging, giving a new perspective to geographical concepts, with man and nature interacting, thus imposing new curricular proposals with regard to school content. With regard to this new paradigm, Cassab (2009, p. 48) states:

> Geography becomes the science of space, a

> space that is inseparable from society. Man, nature and the economy are treated in their multiple interactions and in a dialectical manner. From the 1980s onwards, this geography would influence a series of curricular proposals aimed at the fifth to eighth grades. From then on, the attempt was to bring about a radical change in the content and form of school work. The content is now centred on relations of labour and production. Teaching turns to the study of political, economic and social ideologies and the relationship between society, labour and nature.

Despite the changes that geography has undergone and the attempts at scientific improvement, the content taught at school is far removed from the reality in which the student is inserted, making it difficult for the teacher to work, who in most cases bases his methodology on the textbook. The latter, in turn, does not understand the scenario in which the student finds himself.

Geographical science should enable students to critically analyse their reality, becoming aware of their social rights and responsibilities, so that they can become agents of change in their families, communities, workplaces, schools and wherever they participate (NETO and BARBOSA, 2010).

In the proposal of the National Curriculum Parameters, Geography is a specific area because it offers sufficient support for intervention in society, and through it we can understand the interaction of social agents with nature in order to build their space, understanding the characteristics of the lived place, where bonds are developed, and of other places, different and distant from each other, both in time and space (PCN de Geografia, 1998).

The construction of knowledge is based on the student's reality, and it is up to the teacher to "adapt" the content according to the surrounding environment. According to Bueno and Silva (2011, p. 96), most teaching materials that deal with local space work with specific places, such as São Paulo and Rio de Janeiro. Simpler localities in Brazil are not covered as they should be, leaving the teacher with the task of facilitating their students' understanding of how to create a line of involvement between the contents. The information contained in these materials is complex, as they use very technical language, which makes them difficult to use directly.

> The teaching of geography and the use of didactic and technological resources in primary school from grades 5 to 8 [is] of fundamental importance, because through this subject students can develop their sense of localisation and also understand themselves as part of the world in which they live. Teaching this subject enables the acquisition and improvement of certain concepts that contribute significantly to the development of the student not only as an individual in their environment, but also as a

> citizen in their social environment. These concepts can be utilised in the early grades, as the content covered in geography classes makes it possible to develop both social and physical aspects (CALADO, 2012, p. 13).

According to Guimarães, 1995:

> The teacher has been given the role of building cognitive structures with the student that make them understand themselves and the world, progressively increasing their reflective capacity. The concern must therefore be to work with the student's thinking, combining systematised knowledge and action. In exercising this role in the teaching process, the teacher has for some time been obliged to start from the student's reality, in other words, from everything that is linked to their lived experience (GUIMARÃES, 1995, p. 61).

It is understood that Geography is a strategic discipline in which, initially, the construction of learning is based on the consideration of the reality experienced in everyday life in order to seek various questions that lead the teacher to adequately carry out explanations in the classroom.

Nowadays, the occurrence of difficulties is related to the way in which the didactics and methodologies used in school geography are conducted. Although there are difficult situations faced by teachers, such as low pay, unqualified initial training, excessive working hours, as well as the problem of indiscipline and the absence of the family in the task of educating, teachers must look for alternatives to overcome and transform the reality in which they find themselves (NETO and BARBOSA, 2010).

CHAPTER 3

MATERIALS AND METHODS

In this research, questionnaires with structured questions were administered to the students of the afternoon (90 questionnaires) and evening (45 questionnaires) shifts (model in the appendix) of the Manuel de Matos State School in the municipality of Santana do Mundaú - AL.

The questions were structured to check the students' ability to identify geographical features present in the municipality of Santana do Mundaú-AL and the region.

1.1 Characterisation of the study area

The research was carried out at the Manuel de Matos School in the municipality of Santana do Mundaú - AL, located in the Serrana dos Quilombos Microregion of the state of Alagoas, more precisely in the Vale do Mundaú region (Figure 1).

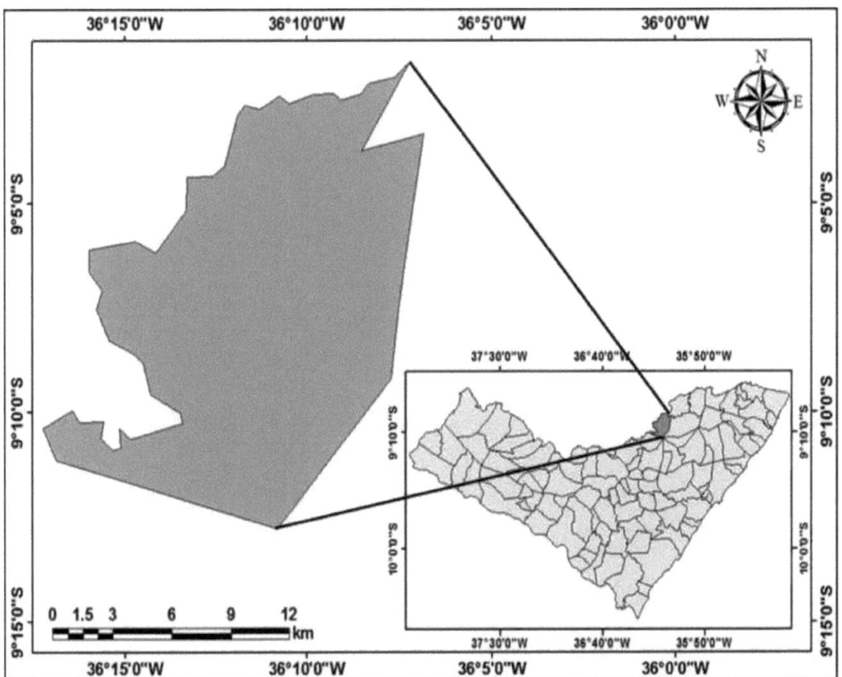

Figura 1. Location of the municipality of Santana do Mundaú in the state of Alagoas. Prepared by: José Thales Pantaleão Ferreira, 2013.

The municipality has a very steep terrain (figure 2), formed mainly by dystrophic Argisols and to a lesser extent by Latosols and Gleissols (EMBRAPA, 2006). The municipality has annual rainfall of around 1,600 mm, starting in February and ending in October (CPRM, 2005).

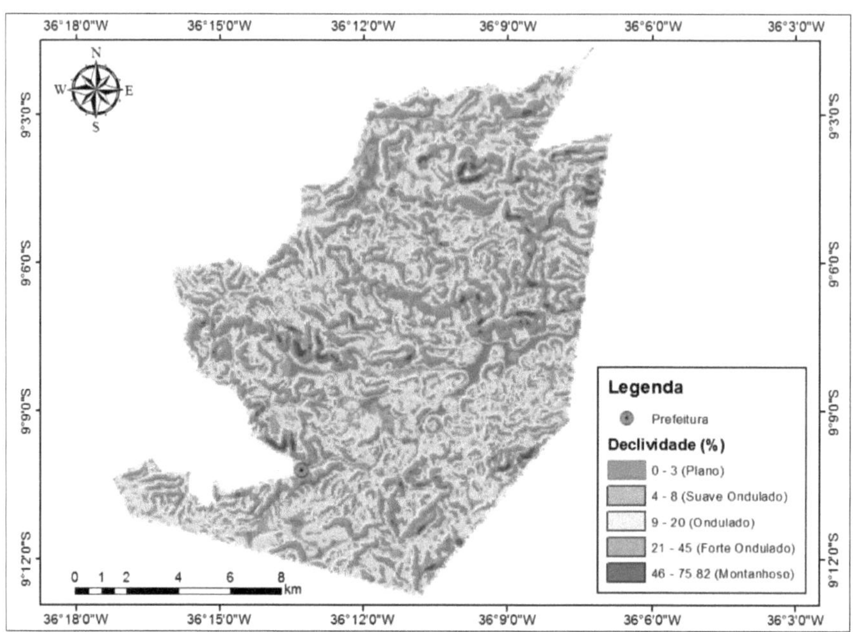

Figura 2. Slope map of the municipality of Santana do Mundaú - AL. Source: Ferreira *et al.* (2012).

The municipality of Santana do Mundaú is located in the middle part of the Mundaú river basin (figures 3 and 4). The Mundaú river basin covers an area of 4,126 km^2 and drains 30 municipalities, half of them in Pernambuco and the other half in Alagoas. The Pernambuco side around 2,155 km^2 while the Alagoas side covers 1,971 km^2 (CARVALHO, 2002).

Figure 3: View from the top of the municipality of Santana do Mundaú. Source: The authors, 2013.

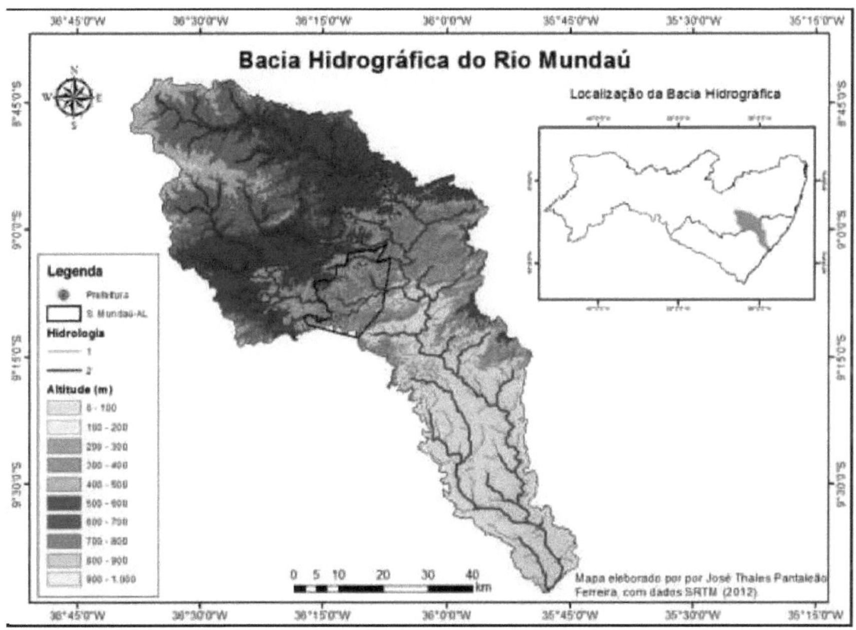

Figure 4. altitude map of the mundaú river basin and the location of the town of Santana do Mundaú-AL. Source: Ferreira *et al.* (2012).

It is a federally-owned river, since its sources are in the

neighbouring state of Pernambuco. In its Pernambuco stretch, the basin is located in the Agreste Pernambucano Mesoregion, where the territories of 15 municipalities are wholly or partially included. With a population of more than 215,705 inhabitants, 8 municipal centres are located in this stretch, the largest of which is the city of Garanhuns. In the Alagoas part of the basin, which corresponds to its lower half, the surface area is 1,971 km^2, where the territories of 15 municipalities in the Mesoregion of Eastern Alagoas are totally or partially inserted.

Covering a population of around 230,000 inhabitants, the Alagoas stretch includes 10 municipal centres, as well as a small part of the urban area of Maceió, with the cities of Rio Largo and União dos Palmares standing out as the main urban centres. This basin lies between the extreme coordinates N and 8,956,000 N; and 780,000 E (25 L) and 174,000 E (24 L) (CARVALHO, 2002).

CHAPTER 4

DIALOGUE WITH THE STUDENTS OF SANTANA DO MUNDAÚ

The Mundaú River passes through the centre of the city of Santana do Mundaú-AL, but when the students were asked which state(s) the Mundaú River basin belongs to with the following alternatives: (a) the State of Alagoas; (b) the State of Pernambuco and Alagoas; (c) the State of Sergipe; (d) the State of Pernambuco; (e) the State of Bahia; only the students on the afternoon shift, the majority of whom (80%) knew the correct answer (b) (Graph 1). The evening shift students were very undecided, with 45% answering that the Mundaú river basin belongs to the states of Pernambuco and Alagoas and another 38% answering option "a" that it only belongs to the state of Alagoas (Graph 1). Some students [afternoon shift (5.6%) and evening shift (13.6%)] also answered option "d" which only belongs to the state of Pernambuco (Graph 1).

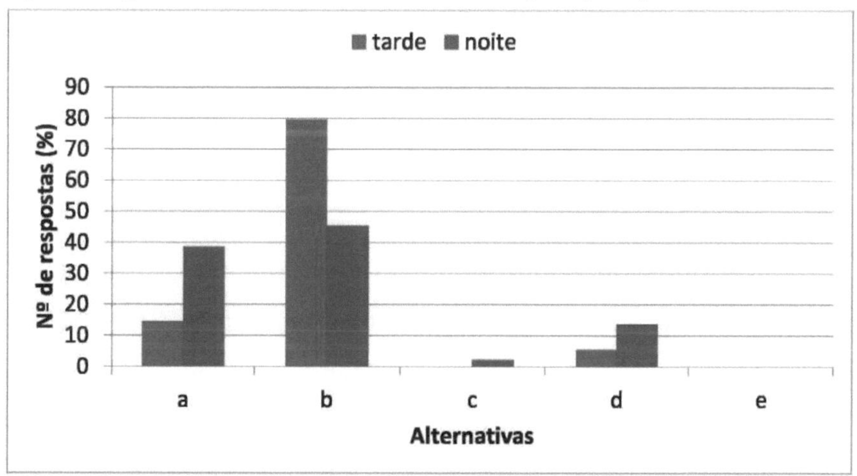

Graph 1. Students' answers to the following question: Which state(s) belong to the Mundaú river basin? Source: The authors,

2013.

The finding that approximately 20% of students on the afternoon shift and 55% on the evening shift (Graph 1) were unaware of basic information about the Mundaú river basin is worrying, as this is important information for students who live in the Mundaú river basin, enjoy its natural resources and are unaware of its spatial location. This river often causes concern about the possibility of flooding during the rainy season, but even so, many students didn't know which state its river basin belonged to.

The town of Santana do Mundaú-AL is approximately 80 kilometres from the coast and within the Atlantic Forest Biome in the state of Alagoas (Figure 5). The Atlantic Rainforest is a tropical forest biome that covers the east, southeast and south coasts of Brazil, eastern Paraguay and the province of Misones in Argentina. Its ecological processes evolved from the Eocene, when the continents were relatively arranged as they are today. The region has been occupied by humans for more than 10,000 years. Since European colonisation, and especially in the 20th century, the Atlantic Rainforest has undergone intense deforestation, with less than 10% of the original vegetation cover remaining (Wikipedia, 2013).

In the question about the natural vegetation of Santana do Mundaú-AL, we asked which type of vegetation is typical of the municipality, with the following alternatives: a) Cerrado; b) Caatinga; c) Mata Atlântica; d) Mata de Araucária; e) Floresta Equatorial. Only 30% of the students on the afternoon shift and 11% on the evening shift got it right [(c) Atlantic Forest] (Graph 2). The vast majority of students from both shifts [afternoon (49%) and evening (66%)] got it wrong and answered that the natural vegetation of Santana do Mundaú is Cerrado (Graph 2).

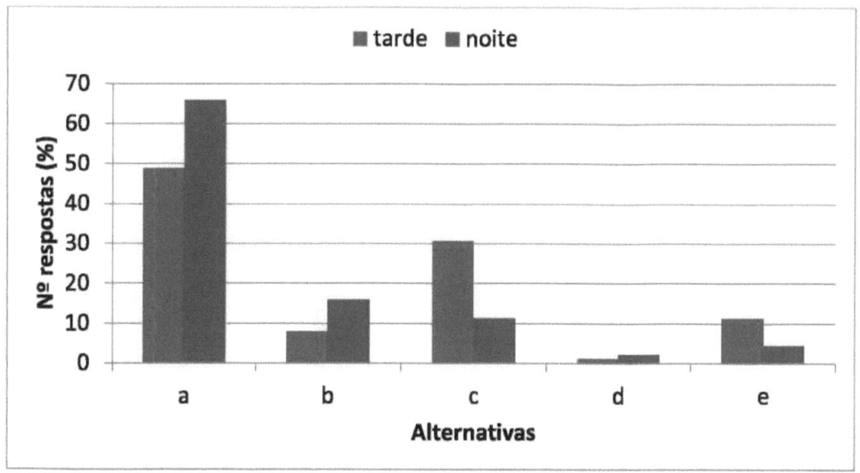

Graph 2. Students' answers to the following question: In terms of natural vegetation, which is typical of Santana do Mundaú? Source: The authors, 2013.

The lack of knowledge about the natural vegetation of their region is worrying, as it is one of the basic pieces of information that students study in Geography, and there may be a lack of connection with what is taught and with the environment that surrounds the students. There may also be an emphasis in textbooks on the Cerrado Biome, since today it is one of the places with the highest agricultural production, following the great devastation of areas of native vegetation.

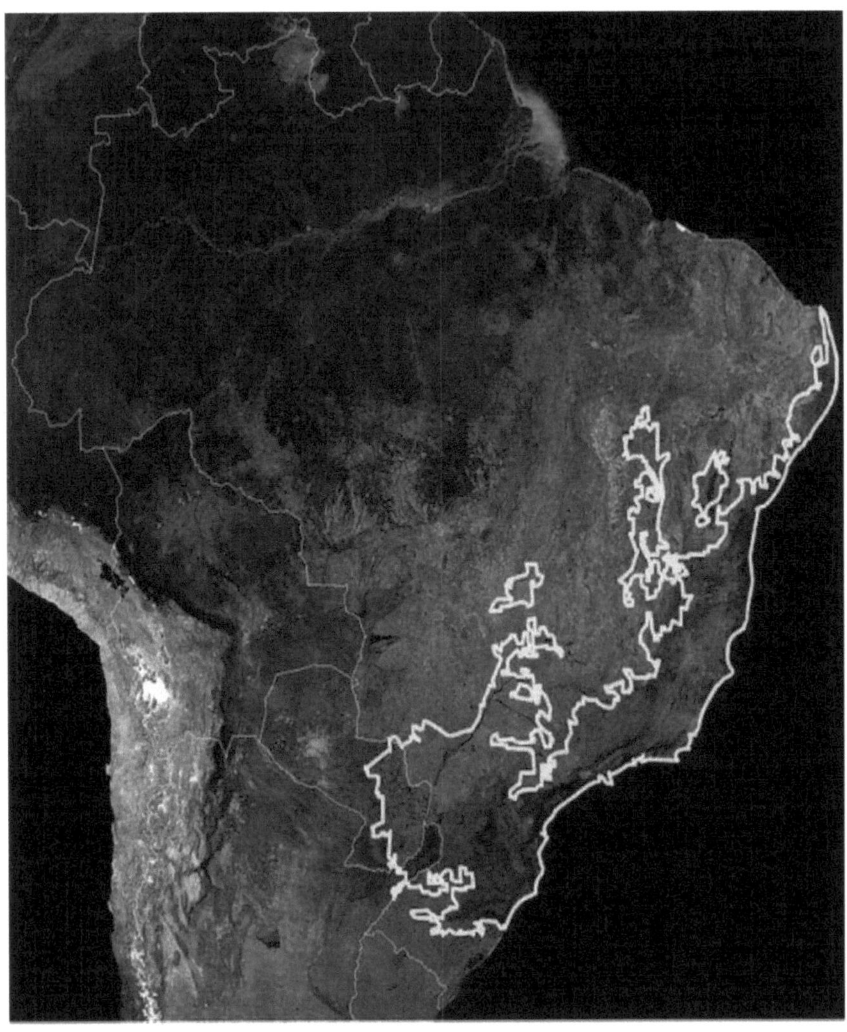

Figure 5: Distribution of the Atlantic Forest Biome in Brazil. Source: Wikipedia (2013).

Watercourses are classified as perennial, intermittent and ephemeral. In perennial watercourses, the sources or springs keep the water flowing all year round. Therefore, the groundwater is responsible for the continuous flow of the rivers. In intermittent streams, the sources or springs are insufficient to maintain the water course throughout the year and in streams classified as ephemeral, there are generally high flows during the rainy seasons and river flow ceases in the dry seasons. In this case, the water table remains above the level of the river flow during the rainy season and below the river bed during the dry season.

The Mundaú River is classified as perennial, maintaining continuous water flow throughout the year.

When the students were asked about the classification of the Mundaú River, they were offered the following alternatives: a) Perennial; b) Intermittent; c) Perennial and intermittent; d) All the alternatives are correct; e) None of the alternatives are correct. 55% of the afternoon shift and only 27% of the evening shift answered option "a" (perennial) (Graph 3).

The evening shift students showed a serious lack of knowledge about the classification of rivers, as the majority (73 per cent) got the question wrong and approximately 41 per cent answered option "d" (all the alternatives are correct), the option that there was no doubt was wrong, as a river cannot have all the classifications (Graph 3). Option "d" was chosen by only 5% of the students on the afternoon shift.

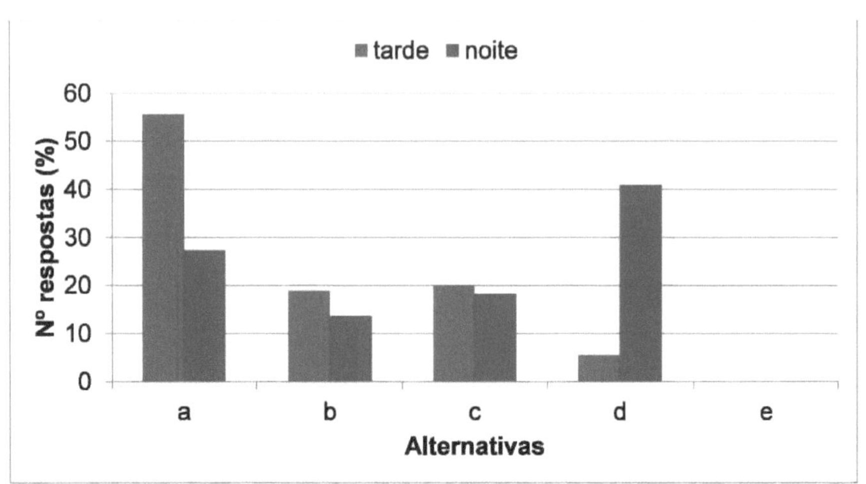

Graph 3. Students' answers to the following question: Regarding hydrography, it can be said that the Mundaú River is:. Source: The authors, 2013.

The lack of knowledge about the classification of rivers and the inability to know the classification of the main river in their city are worrying, showing that there is no efficient assimilation of the content taught in the subject of geography, as well as possibly not taking advantage of the natural wealth of the region to exemplify the subjects taught. This unpreparedness is worrying and could have negative effects not only in everyday life, but also in exams and entrance exams.

The city of Santana do Mundaú is located in the mesoregion of eastern Alagoas and the micro-region of the quilombos (SEPLANDE, 2013). This is very important information, especially for people from Alagoas who want to travel in the state and get to know their state. However, when the students were asked in which mesoregion and microregion the municipality of Santana do Mundaú is located, they were given the following alternatives: (a) Sertão alagoano and mata alagoana; (b) Agreste and litoral norte alagoano; (c) Leste alagoano and serrana dos quilombos; (d) Sertão alagoano and serrana dos quilombos; (e) None of the alternatives; only 21% and 18%, respectively from the afternoon and evening shifts answered the correct option [(c) leste alagoano and serrana dos quilombos] (Graph 4). The remaining 79% (afternoon shift) and 82% (night shift) answered other options, with option "a" (sertão alagoano and mata alagoana) standing out with 30% and 29%, respectively from the afternoon and night shifts (Graph 4).

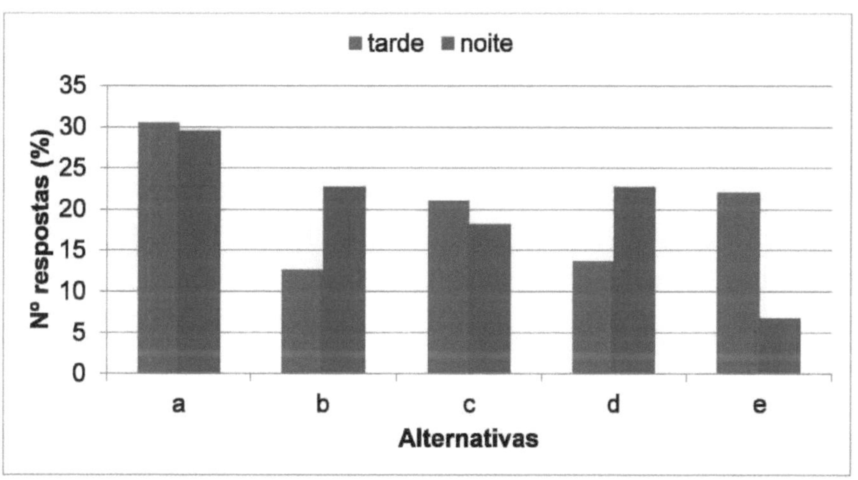

Graph 4. Students' answers to the following question: Which mesoregion and microregion is the municipality of Santana do Mundaú located in? Source: The authors, 2013.

The choice of option "a" (Graph 4) by the students may have been influenced by the common use of the location of Santana do

Mundaú as the Alagoas Mata Zone, which is not wrong, but is academically better known as the mesoregion of eastern Alagoas and the mountain micro-region of the quilombos, but alternative "a" also included Sertão Alagoas, which is a long way away and has totally different soil and climate characteristics.

The municipality of União dos Palmares borders Santana do Mundaú, the two being linked by the AL 101 motorway heading towards the state capital. Santana do Mundaú has a predominantly undulating terrain (9 to 20% slope) and a strongly undulating terrain (21 to 45% slope) (Figure 2), unlike União dos Palmares, which has a more flattened terrain (gently undulating to undulating).

When the students were asked if there were any differences between the reliefs of União dos Palmares and Santana do Mundaú, the majority of students, 88% and 73%, from the afternoon and evening shifts respectively, replied that there were differences in the reliefs of the municipalities. This result shows that the students are able to read the geomorphology of their region when they realise that there is a difference with the neighbouring town. However, it is still serious that 27% of the evening shift students were unable to perceive this clear difference in relief between the municipalities. (Graph 05)

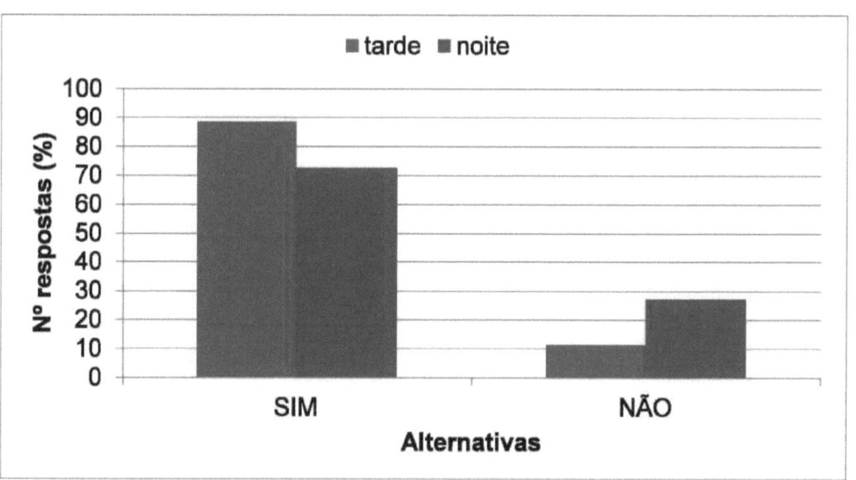

Graph 5. Students' answers to the following question: Do you notice any differences between the relief of União dos Palmares and Santana do Mundaú? Source: The authors, 2013.

The students were asked which relief classification, according to Jurandyr Ross, the municipality of Santana do Mundaú would fall under, and were offered the following alternatives: a) Sertaneja and São Francisco Depression; b) Borborema Plateau; c) Coastal plains and tablelands; d) Tocantins Depression; e) None of the alternatives; however, only 15% and 16%, respectively from the afternoon and evening shifts, answered correctly (Graph 6). It is noteworthy that 53% of the afternoon students thought that none of the alternatives were correct ("e" option) and

also that 36 per cent of the evening shift students chose option "d" (depression of the Tocantins), it was not expected that the students would err in this way, even choosing depression of the Tocantins.

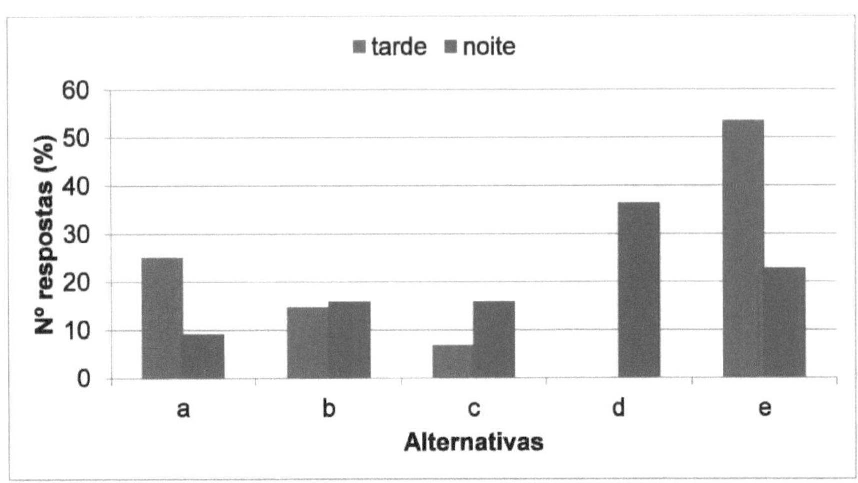

Graph 6. Students' answers to the following question: According to Jurandir Ross, there are three forms of relief. Which of the following classifications does Santana do Mundaú belong to?

According to the division of Brazil's relief proposed by Professor Jurandyr Ross, which is an improvement on Professor Ab'Saber's division, the municipality of Santana do Mundaú is part of the Borborema Plateau, which covers the states of Alagoas, Pernambuco, Paraíba and Rio Grande do Norte. As shown in Figure 6, Santana do Mundaú is part of the Borborema Plateau.

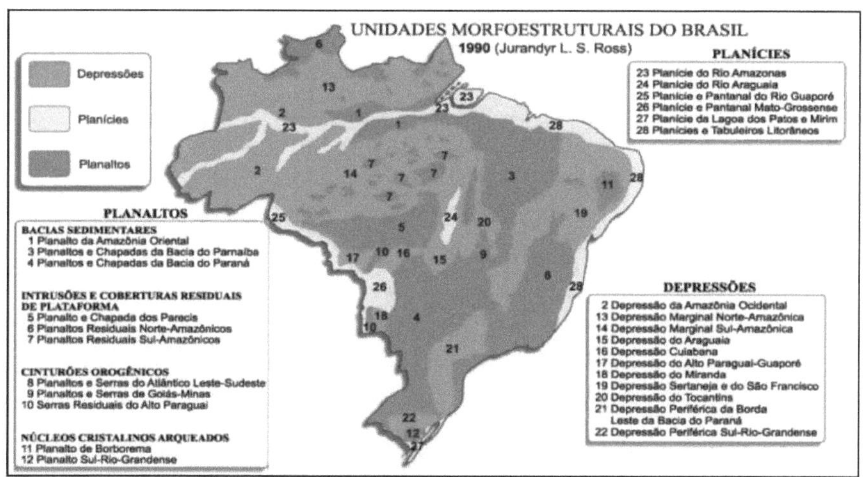

Figure 6. Morphostructural units of Brazil according to the classification of Jurandyr L. S. Ross.

The students from both shifts (afternoon and evening) were able to successfully identify the occupation of the physical space of the municipality of Santana do Mundaú-AL, through the alternatives: (a) banana plantation and pine cone; (b) sugar cane plantation and pineapple; (c) lime orange plantation, pasture and sugar cane; (d) pasture, banana plantation and lime orange plantation; (e) all alternatives; 68% and 58% got it right when answering that the physical space of their municipality is occupied mainly with pasture, banana plantation and lime orange plantation (Graph 7). However, a significant number of evening shift students (35%) and afternoon shift students (27%) got it wrong and answered option "c" lime orange plantation, pasture and sugar cane (Graph 7). However, this is possibly due to the influence of sugar cane cultivation in the state of Alagoas, leading the students to think that this crop is also very important in Santana do Mundaú, but the local terrain must have made it difficult to plant large areas of sugar cane in Santana do Mundaú and given the opportunity to other agricultural crops.

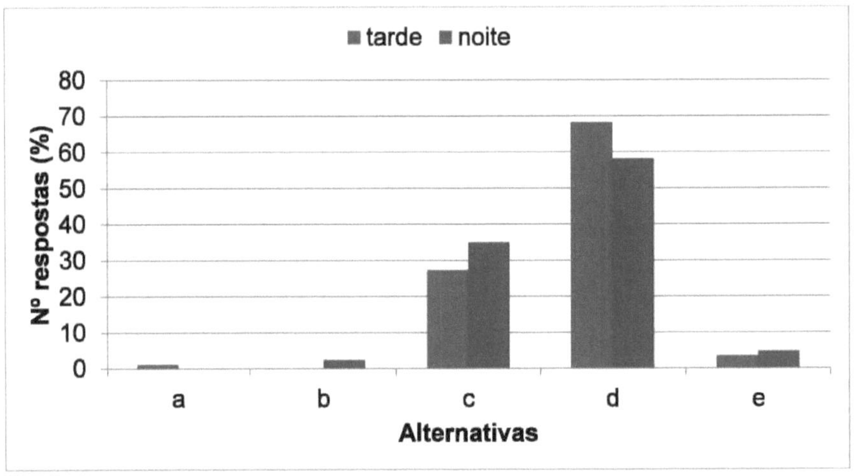

Graph 7. Students' answers to the following question: Santana do Mundaú is surrounded by mountains. Its physical space is mainly occupied by:

In Santana do Mundaú, planting without contour lines is common, popularly known as "downhill", as illustrated in figure 6, facilitating soil erosion and degradation (FERREIRA *et al.,* 2012). Contour lines are lines that represent the irregularities of the relief. They are widely used to represent the relief on topographic maps.

This procedure of planting "downhill" in rainy seasons has led to problems of slope instability in specific places in the municipality, with the occurrence of landslides and mudflows, mobilising fractured blocks and silting up rivers and lagoons (FERREIRA et al., 2012).

According to SANTOS FILHO *et a!,* (2005) this type of planting was common in the past, but today this practice is unacceptable, and when it happens it reflects the inoperativeness of the technical assistance and rural extension services of local organisations, adding that this planting procedure can cause a generalised problem of environmental balance.

In order to prevent erosion of these soils in the topographical conditions prevailing in the municipality, more sustainable agricultural management should be adopted, with level-curve planting, the use of terracing and always maintaining a vegetation cover on the ground, preventing rainwater drops from falling directly on the soil and being washed away, removing the most fertile layer of soil and consequently reducing agricultural production (FERREIRA et al., 2012).

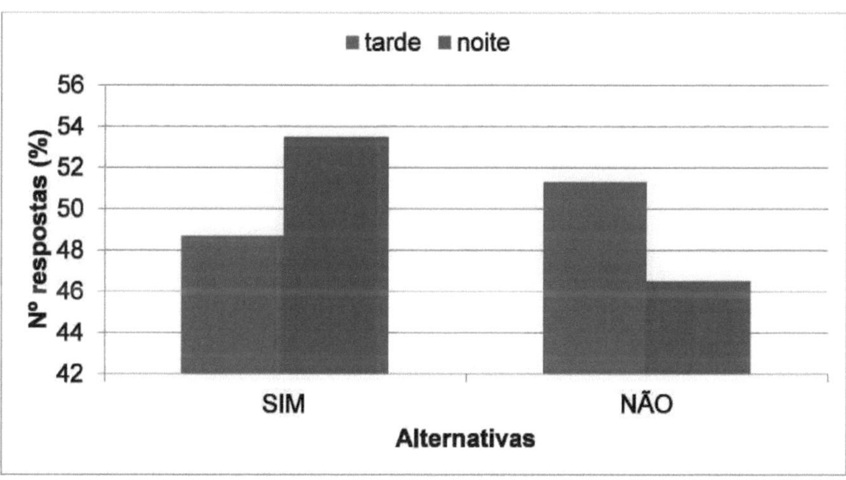

Graph 8. Students' answers to the following question: Understanding contour lines is of great importance for agricultural production, as it can prevent many erosion processes. Do you think that agricultural crops in Santana do Mundaú are planted on contour lines?

When the students were asked about the use of contour lines

in agricultural plantations, 53% and 49%, respectively from the evening and afternoon shifts, were wrong to say that contour lines are used. However, 51% of the afternoon shift and 47% of the night shift answered that the crops are not planted on a contour line, which is the correct alternative. It is clear that these students, despite basically living in rural areas, as the town is surrounded by orange plantations, do not know how to identify the way in which this agricultural crop is planted. This result is worrying, as the technique of planting on a contour line is one of the ways to prevent soil erosion, especially in Santana do Mundaú, which has undulating to strongly undulating terrain (FERREIRA *et al.,* 2012).

Figura 7. Planting lime oranges without respecting contour lines in Santana do Mundaú - AL. Source: Ferreira et al. (2012).

Agricultural soils can be affected by erosive agents. The students were asked which erosive agent has the greatest impact on Santana do Mundaú, offering the following alternatives: a) Pluvial (rain); b) Fluvial (rivers); c) Wind (winds); d) Marine (sea).

If you look at Graph 9 with the answers, you can see that the students in the afternoon were the ones who got the main erosive agent of Santana's soils right, with 55 per cent in the afternoon compared to 31 per cent in the evening.

The students' doubts about the reason for soil erosion cannot be overlooked, when 36% in the afternoon and 63% in the evening answered fluvial erosion. However, when analysing agricultural regions, it can be seen that there are a number of forms of erosion, but the one that predominates is pluvial erosion, due to the misuse of the soil, which leaves it unprotected, and in the first rains drags a lot of soil into rivers and streams, thus causing two environmental problems, the loss of fertile soil and the silting up of rivers.

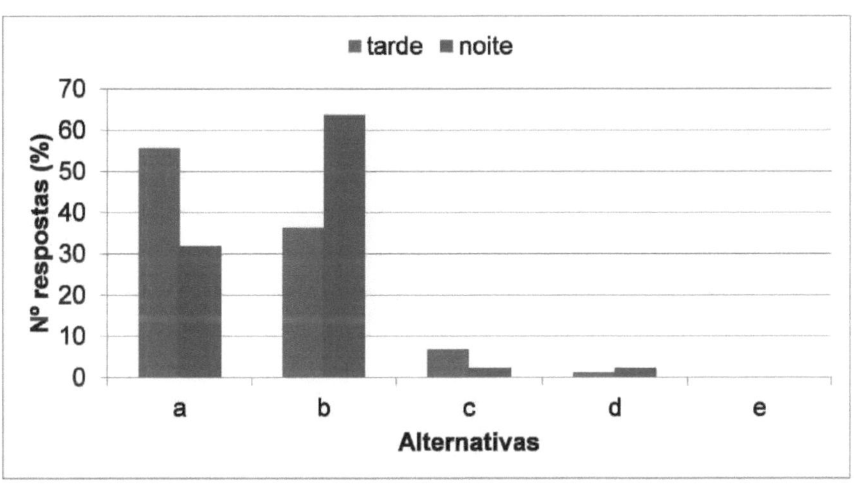

Graph 9. Question 9.) Agricultural soils can be affected by erosive agents. In Santana do Mundaú, which erosive agent has the greatest impact?

We asked the students about the municipality's climate,

proposing the following choices: a) Tropical hot; b) Equatorial; c) Tropical rainy with dry summers; d) Continental dry. Analysing the graph of answers to question 10, it can be concluded that 58% of the students in the period

in the afternoon ticked the correct alternative, compared to 56% in the evening, where they ticked alternative (c) (Tropical rainy with dry summer). However, many students ticked alternative (a) in the afternoon (24%) and the evening (31%), and this answer does not match the climate of the city studied, showing once again the doubt expressed by the students.

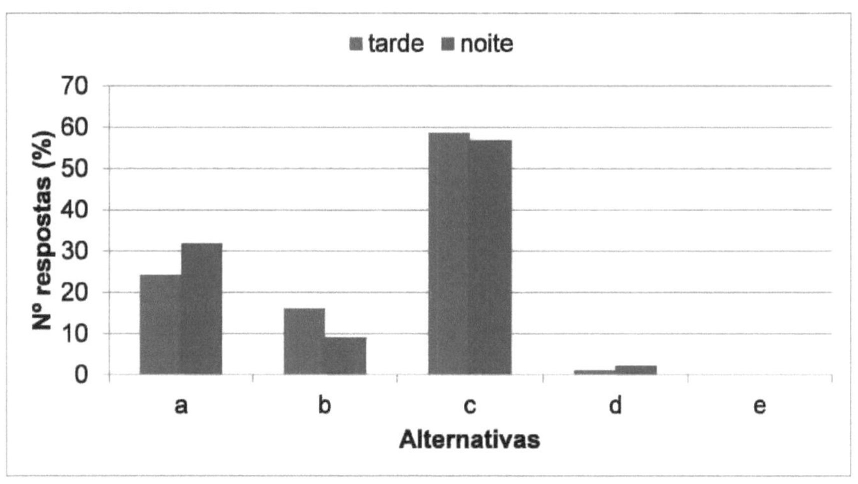

Graph 10. Question 10) With regard to climate, it can be said that the climate of the municipality of Santana do Mundaú is:

The average annual rainfall in Santana do Mundaú is 1600 mm, which can be considered a region with a high rate of precipitation compared to the semi-arid regions of Brazil.

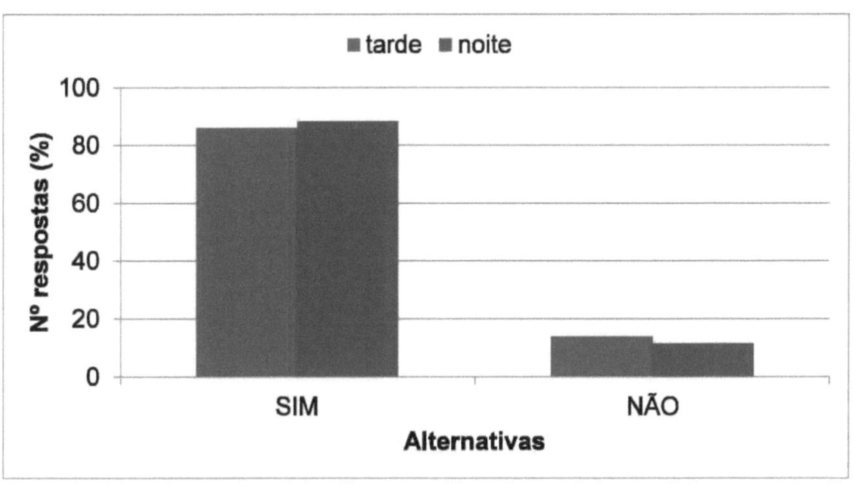

Graph 11 - Question 11 - Do you like the subject of Geography?

As can be seen in the graph representing question 11, 86 per cent of the afternoon and 88 per cent of the evening students answered that they liked the subject of geography. However, they found it difficult to answer the questions, which contextualise the content seen in the classroom with local environmental characteristics.

This shows that there is something wrong with the learning of geography in this school, or perhaps that the students are unable to associate what they have learnt in the classroom with everyday life.

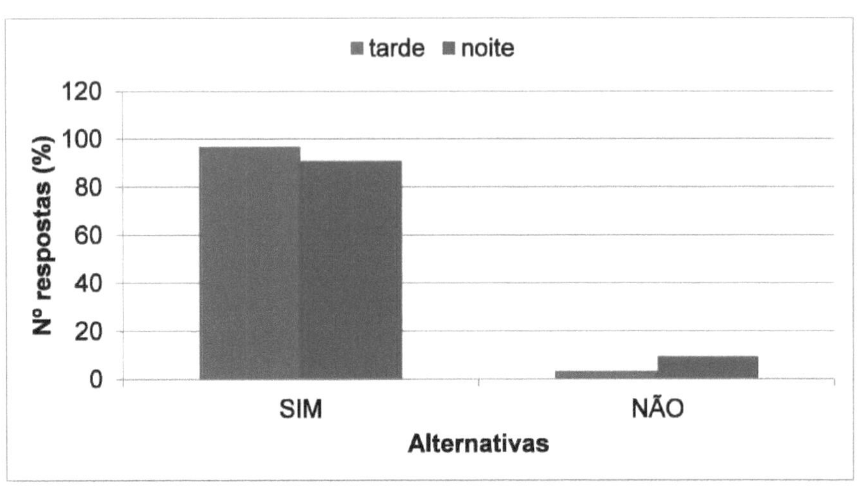

Graph 12 Question 12: Have you learnt anything that could help you throughout your life in Geography?

The graph for question 12 shows that 96 per cent of students in the afternoon and 90 per cent in the evening answered that Geography will help them throughout their lives. This shows that the students have a good relationship with the subject of Geography and are ready to learn, which can facilitate the teaching-learning process.

When the students were asked what they thought of the subject of Geography: a) interesting; b) tiring; c) not very interesting; d) I'm not interested (graph 13), 85% of the afternoon and 84% of the evening students replied that the lessons were interesting. This raises the question: How can such interesting lessons not produce the expected results when taken from theory to practice?

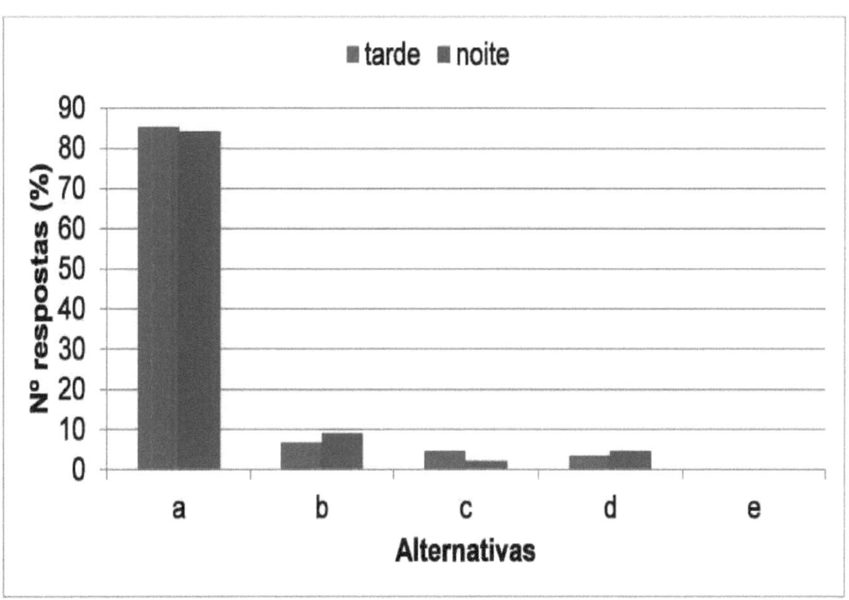

Graph 13. Question 13: What do you think Geography lessons are like?

Finally, the students were asked: If you could choose, would you have geography lessons? 93% of the evening students said yes. However, 50 per cent of the evening students said yes, compared to 50 per cent no. As can be seen in Graph 14, there is a big difference between the responses of the different shifts, with evening students being more accepting of geography, while afternoon students remain divided.

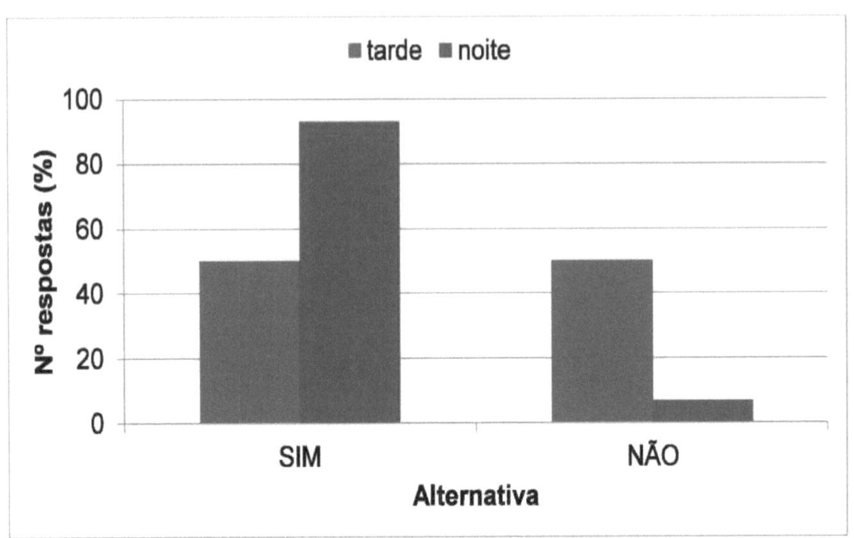

Graph 14. Question 14) If you could choose, would you take a geography class?

This finding differs from question 13 (graph 13), in which more than 80 per cent of both shifts said that they found the classes interesting, but in this question 13 many said that if they could they wouldn't want to take geography classes. This result is worrying, because the subject of geography allows students to critically analyse their reality, becoming aware of their social rights and responsibilities, so that they can become agents of change in their family, community, work, school, among other social environments (NETO and BARBOSA, 2010).

CHAPTER 5

FINAL CONSIDERATIONS

Geography, as a science that deals with the broad issues facing society in general, must act as an intermediary between the knowledge offered to students, in a way that helps them to be critical and thinking. But for this to really happen, it is necessary to leave behind some of the practices of the past, which do not contribute satisfactorily to bringing about change. During the course of the research, the students found it very difficult to solve the questions in the questionnaire. In particular, students on the afternoon shift were more successful in their answers than those on the evening shift. This proves that there is a clear difference in the form of teaching and learning applied to both groups of students. It is inferred that the fact that the evening shift offers a lower workload affects student performance, as well as physical and mental tiredness.

A significant number of students, mainly from the evening shift, were unable to identify geographical features present in their municipality and region, while those from the afternoon shift did much better.

Overall, it can be seen that there is still a gap between the reconciliation of general geography knowledge with the local environment on the part of teachers in the municipality of Santana do Mundaú/AL, and that there is an immediate need to change the methodologies used. It is up to each professional to self-analyse their own practices and take the initiative to change them.

It was realised that students need to interact more with the subject, so that it stops being exclusively classroom-based and starts having a broader view of its surroundings, given that this municipality provides this possibility. Out-of-class activities can

provides an excellent opportunity to create a link between the areas, becoming an auxiliary channel for teachers, who according to the research carried out, still don't carry it out as standard, but only sporadically. In this way, as well as being able to understand the environment in which they live, students also become aware of and understand the changes that can occur in that environment. An example of a very common change in Santana do Mundaú is the silting up of the river that cuts through the city, the Rio Mundaú, which has been suffering from this problem as a result of the accumulation of rubbish deposited by the residents themselves and the removal of the riparian forest to build properties and plant crops. Addressing issues like this would result in much more productive lessons and more interested and critical students, not to mention the results, which could have repercussions in new actions as a result of this awareness. These are issues that can be tackled both inside and outside school. There is no shortage of environmental problems in Santana do Mundaú that can be addressed in class. They can be used to attract students to mobilise in favour of the causes that permeate against the removal of the remaining riparian forest, creating projects in conjunction with the competent bodies, holding seminars on the consequences of siltation and the reasons that cause it, the main one being rain erosion, which triggers another, which is soil infertility, bringing agricultural losses to the municipality and consequently to these students, most of whom are the children of small farmers, who make their living from the soil, which they are helping to destroy due to a lack of information, and the school can help to minimise this problem.

However, it should be made clear that this is not the school's responsibility, but that of the municipality's agriculture department. However, this does not prevent it from playing this role, which can contribute both to the student's learning and to the way in which the soil is used and handled, through lectures for students and even their parents on erosion and its environmental and economic consequences, which can be given by the students themselves or by a qualified professional that the municipality has at its disposal.

Thus, there needs to be a change in the teaching practices used by teachers, especially in primary schools, where the predominant use of textbooks is the primary means of transmitting knowledge, without taking into account the real priorities for the student's intellectual growth. This reality characterises the teaching of geography in the state of Alagoas, where there are still very rudimentary ways of presenting the subject, especially in state schools.

This research was just the start of a series of questions that still need to be unravelled and worked on in the educational sphere of the Manoel de Matos State School, especially with regard to the students' latest answers, in which they say they like the subject of Geography, but would choose not to have it in their school curriculum. Despite claiming to have learnt aspects of Geography that they will carry with them throughout their lives, they were unable to get simple questions right, some of which were based on pure observation. There are still many gaps and questions within the classroom environment at this school regarding the teaching of local geography, but we hope that these themes will continue to be explored in greater depth along the lines already established in this work, as well as in various other areas that encourage students to get to know their municipality.

CHAPTER 6

BIBLIOGRAPHICAL REFERENCES

BUENO, Míriam Aparecida. SILVA, Karine Araújo. Analysing the teaching and learning of local space and the training of primary school teachers in public schools in the metropolitan region of Goiânia (RMG). R. Ens. Geogr., Uberlândia, v. 2, n. 3, p. 95-112, jul./dez. 2011. ISSN 2179-4510.

CALADO, Flaviana Moreira. Geography teaching and the use of didactic and technological resources. Geosaberes, Fortaleza, v. 3, n. 5, p. 12-20, jan. / jun. 2012. © 2010, Federal University of Ceará.

CARVALHO, G.S. de. Potential water resources of the mundaú river basin. In: Hydrological studies of the mundaú river basin. State Secretariat for the Environment and Water Resources. Maceió-AL, 2002, 67p. Available at: < http://www.semarh.al.gov.br/programas/arquivos-para-baixar/Estudos%20Hidrologicos%20Rio%20Mundau.zip/view.>, accessed on 31/08/2013

CASSAB, Clarice. REFLECTIONS ON THE TEACHING OF GEOGRAPHY. Geografia: Ensino & Pesquisa, Santa Maria, v. 13 n. 1, p. 43 50, 2009.

FERREIRA, E.P.; FERREIRA, J.T.P.; PANTALEÃO, F.S.; ALBUQUERQUE, K.N. FERREIRA, A.C. Citriculture in Santana do Mundaú AL: agricultural management of the lime orange Citrus sinensis (L.) Osbeck and the challenges for the sustainability of the crop. Enciclopédia biosfera, v. 8, p. 203-219, 2012.

GUIMARÃES, Iara Vieira. Geography teaching in times of globalisation and paradigmatic crisis. Ensino e revista, 4 (1): 59-64, Jan./Dec. 1995.

IBGE - Brazilian Institute of Geography and Statistics. Santana do Mundaú- AL. Available at: <http://www.ibqe.qov.br/cidadesat/painel/painel.php?codmun=270810>, accessed on 31/08/2013.

NETO, Francisco Otávio Landim. BARBOSA, Maria Edivani Silva. GEOGRAPHY TEACHING IN BASIC EDUCATION: an analysis of the relationship between teacher training and their performance in school geography. Geosaberes - v. 1, n. 2, December/2010.

OLIVEIRA, Antônio Marcos Machado de; MIRANDA, Sérgio Luiz. The importance of teaching Geography today. R. Ens. Geogr., Uberlândia, v. 1, n. 1, p. 1-2, Jul./Dec. 2010.

Parâmetros curriculares nacionais: geografia /Secretaria de

Educação Fundamental. Brasília: MEC/SEF, 1998. P.15. Available at: http://portal.mec.qov.br/seb/arquivos/pdf/qeoqrafia.pdf, accessed on 31/08/2013.

SANTOS FILHO, H. P; MAGALHÃES, A. F. J.; COELHO, Y. S. (editors). Citrus: The producer asks, Embrapa answers. Brasília - DF: Embrapa Informações Tecnológicas, 219 p.: il. 2005.

SEPLANDE - State Secretariat for Planning and Economic Development. Municipal Profile: Santana do Mundaú. Maceió-AL, n.1, 27p. 2013. Available at: < http://informacao.seplande.al.qov.br/perfil-municipal/relatorios/Municipal Santana%20do%20Munda%FA 2O12.pdf>, accessed on 31/08/2013.

WIKIPEDIA. Atlantic Rainforest. Available at: < http://pt.wikipedia.org/wiki/Mata Atl%C3%A2ntica>, accessed on 31/08/2013.

CHAPTER 7

APPENDIX

Questionnaire applied

Answer the questionnaire below

Class: Shift:

1) To which state(s) does the Mundaú river basin belong?

a) To the State of Alagoas;
b) The states of Pernambuco and Alagoas;
c) To the State of Sergipe;
d) To the State of Pernambuco;
e) The State of Bahia.

2) What natural vegetation is typical of Santana do Mundaú?

a) Cerrado;
b) Caatinga;
c) Atlantic Forest;
d) Araucaria forest.
e) Equatorial forest.

3) In terms of hydrography, the Mundaú River is:
a) Perennial;
b) Intermittent;
c) Perennial and intermittent;
d) All the alternatives are correct.

4) Which mesoregion and microregion is the municipality of Santana do Mundaú located in?

a) The Alagoas hinterland and the Alagoas forest;
b) Agreste and north coast of Alagoas;
c) East of Alagoas and the quilombos;
d) The Alagoas hinterland and the quilombo mountain range;
e) None of the alternatives

5) Do you notice any differences between the terrain of União dos Palmares and Santana do Mundaú?

( ) YES
( ) NO

6) According to Jurandir Ross, there are three forms of relief. Which of the following classifications does Santana do Mundaú fall into?

a) Sertaneja and São Francisco Depression;
b) Borborema Plateau;
c) Coastal plains and tablelands;
d) Tocantins Depression;
e) None of the above.

7) Santana do Mundaú is surrounded by mountains. Its physical space is mainly occupied by:

a) Banana and pine cone plantations;
b) Sugarcane and pineapple plantations;

c) Lime orange plantation, pasture and sugar cane;

d) Pastures, banana plantations and lime orange plantations;

e) All alternatives.

8) Understanding contour lines is of great importance for agricultural production, as it can prevent many erosion processes. Do you think that agricultural crops in Santana do Mundaú are planted on contour lines?

( ) YES

( )NO

9) Agricultural soils can be affected by erosive agents. In Santana do Mundaú, which erosive agent has the greatest impact?

a) Rainfall;

b) Fluvial (rivers);

c) Wind;

d) Navy (sea).

10) The climate of the municipality of Santana do Mundaú is as follows:

a) Tropical hot;

b) Equatorial;

c) Tropical rainy season with dry summers;

d) Dry continental.

11) Do you like Geography?

( ) NO

( ) YES

12) Have you learnt anything that can help you throughout your life in Geography?

( ) YES

( ) NO

13) Geography lessons are for you:

a) INTERESTING
b) CANSATIVES
c) FEW INTERESTING (Insert)
d) I'M NOT INTERESTED

14) If you could choose, would you take a geography class?

( ) YES

( ) NO

Printed by Books on Demand GmbH, Norderstedt / Germany